Visit Susie and her Animal Friends at
KidsKyngdom.com

For my 3 Children

CAMILET

That's a mix up of
Cameron, Dominic & Violet

Thank You SO Much for
all of Your Support!

Visit Susie and her Animal Friends at
KidsKyngdom.com

Visit Susie and her Animal Friends at
KidsKyngdom.com

Thank you for supporting
Kids Kyngdom

Get Your **FREE**
audio book
at
KidsKyngdom.com

Send Dr. Susie a Creative Message on Instagram

Instagram

I may pick You!

Hi Kyngs!
You found us!

I'm with my good friend

Bash

He's a frog

We are in
South America today

We are visiting an animal who likes to eat **ants**

That's right!

We're visiting

Anteaters!

Visit Susie and her Animal Friends at
KidsKyngdom.com

They **LOVE** Ants!

Not pants!

Ants!

Especially ants

that dance

Guess how many **teeth**

an anteater has?

That's right! Zero!

Not hero!

Zero

0

Luckily they have super

sticky tongues

Visit Susie and her Animal Friends at
KidsKyngdom.com

Their sticky tongue

helps them to eat

They use their claws to find around 30,000 Ants every day!

Anteaters have a great sense of smell

They can smell 40 times better than us

Not sporty

Visit Susie and her Animal Friends at
KidsKyngdom.com

Forty!

Baby anteaters travel on

their Mom's back

They hang on tight

for 1 full year

Their Moms teach them

how to hunt ants for 3 years

Visit Susie and her Animal Friends at
KidsKyngdom.com

Then they start hunting ants

all by themselves

Come on Kyngs!

And Anteaters!

Visit Susie and her Animal Friends at
KidsKyngdom.com

Let's find another animal with a really long nose

Not hose

Nose!

That's right!

Let's go visit some elephants!

Visit Susie and her Animal Friends at
KidsKyngdom.com

ISBN: 978-1-959501-02-2
Published by: Kyngdom, LLC

What is Sammie Kyng's Favorite Animal?

Sammie **Kyng's** favorite animal used to be an **Anteater.**

Now it's an **Elephant**!

If you have a favorite animal you would like Doctor Susie to visit, just **send a request** to Kidskyngdom.com or Instagram.

Kidskyngdom.Com

Instagram

Visit Susie and her Animal Friends at
KidsKyngdom.com

Did you **LOVE** learning about **Anteaters**

with Susie and Bash?

Please share your review

Visit Susie and her Animal Friends at
KidsKyngdom.com

www.ingramcontent.com/pod-product-compliance
Lightning Source LLC
Chambersburg PA
CBHW051347290326
41933CB00042B/3322